WAVES

THINGS YOU SHOULD KNOW
(QUESTIONS AND ANSWERS)

By Rumi Michael Leigh

Introduction

I would like to thank you for purchasing this book, *"Waves, things you should know (questions and answers)"*.

This book will help you understand, revise, and have a good general knowledge and understanding of the basics of wave exercises.

I hope you enjoy it!

Table of Contents

Part 1: Waves

Exercise 1

Questions

a) A guitar string has a frequency of 440 Hz. What is the wavelength of the sound wave produced by the string if the speed of sound is 340 m/s?

b) A water wave has a frequency of 5 Hz and a wavelength of 2 m. What is the speed of the wave?

c) An earthquake produces a P-wave with a speed of 6 km/s and an S-wave with a speed of 3.5 km/s. If the distance from the earthquake epicenter to a seismograph station is 1000 km, how long does it take for the P-wave and S-wave to reach the station?

d) A violin string is 30 cm long and has a frequency of 440 Hz. What is the speed of the wave on the string?

e) A radio station broadcasts at a frequency of 92.5 MHz. What is the wavelength of the radio wave if the speed of light is 3×10^8 m/s?

Answers

a) A guitar string has a frequency of 440 Hz. What is the wavelength of the sound wave produced by the string if the speed of sound is 340 m/s?

(f) = 440 Hz and the speed of sound (v) = 340 m/s.

$v = f\lambda$, where λ is the wavelength.

$\lambda = v/f = 340/440 = 0.773$ m.

Answer: the wavelength of the sound wave produced by the guitar string is 0.773 m.

b) A water wave has a frequency of 5 Hz and a wavelength of 2 m. What is the speed of the wave?

(f) = 5 Hz

wavelength (λ) = 2 m.

$v = f\lambda$, where v is the speed of the wave.

$v = 5 \times 2 = 10$ m/s.

Answer: the speed of the water wave is 10 m/s.

c) An earthquake produces a P-wave with a speed of 6 km/s and an S-wave with a speed of 3.5 km/s. If the distance from the earthquake epicenter to a seismograph station is 1000 km, how long does it take for the P-wave and S-wave to reach the station?

$t = d/v$, where t is the time taken, d is the distance, and v is the speed.

For the P-wave, $t = 1000/6 = 166.67$ s.

For the S-wave, $t = 1000/3.5 = 285.71$ s.

Answer: P-wave takes 166.67 s and the S-wave takes 285.71 s to reach the seismograph station.

d) A violin string is 30 cm long and has a frequency of 440 Hz. What is the speed of the wave on the string?

$v = f\lambda$,

where v is the speed of the wave

f is the frequency

λ is the wavelength

$\lambda = 2L$, where L is the length of the string.

$\lambda = 2 \times 0.3$ m $= 0.6$ m.

$v = f\lambda = 440 \times 0.6 = 264$ m/s.

Answer: the speed of the wave on the violin string is 264 m/s.

e) A radio station broadcasts at a frequency of 92.5 MHz. What is the wavelength of the radio wave if the speed of light is 3 x 10^8 m/s?

$v = f\lambda$, where v is the speed of the wave, f is the frequency, and λ is the wavelength.

$\lambda = v/f$

$\lambda = 3 \times 10^8 / 92.5 \times 10^6 = 3.243$ m.

Answer: the wavelength of the radio wave is 3.243 m.

Exercise 2

Questions

a) An ocean wave has a frequency of 0.2 Hz and a wavelength of 20 m. What is the period of the wave?

b) A sound wave travels through air with a frequency of 1000 Hz and a wavelength of 34 cm. What is the distance the wave travels in 2 seconds?

$v = f\lambda$, where v is the speed of the wave, f is the frequency, and λ is the wavelength.

c) A light wave has a frequency of 5×10^{14} Hz and a wavelength of 600 nm. What is the energy of the wave in electronvolts (eV)?

d) A guitar string is plucked and produces a standing wave with two nodes and three antinodes. If the length of the string is 60 cm, what is the wavelength of the wave?

Answers

a) An ocean wave has a frequency of 0.2 Hz and a wavelength of 20 m. What is the period of the wave?

$T = 1/f$, where T is the period and f is the frequency

$T = 1/0.2 = 5$ s.

Answer: the period of the ocean wave is 5 seconds.

b) A sound wave travels through air with a frequency of 1000 Hz and a wavelength of 34 cm. What is the distance the wave travels in 2 seconds?

$v = f\lambda$, where v is the speed of the wave, f is the frequency, and λ is the wavelength.

$v = f\lambda$

$v = 1000 \times 0.34 = 340$ m/s

$d = vt$, where d is the distance and t is the time.

$d = 340 \times 2 = 680$ m.

Answer: the sound wave travels a distance of 680 meters in 2 seconds.

c) A light wave has a frequency of 5×10^{14} Hz and a wavelength of 600 nm. What is the energy of the wave in electronvolts (eV)?

$E = hf$, where E is the energy of the wave, h is Planck's constant (6.626×10^{-34} J.s), and f is the frequency.

$E = hf$

$E = 6.626 \times 10^{-34} \times 5 \times 10^{14} = 3.313 \times 10^{-19}$ J.

To convert the energy to electronvolts, we can use the conversion factor 1 eV = 1.602×10^{-19} J.

E = (3.313 x 10^-19)/(1.602 x 10^-19) = 2.07 eV.

Answer: the energy of the light wave is 2.07 electronvolts.

d) A guitar string is plucked and produces a standing wave with two nodes and three antinodes. If the length of the string is 60 cm, what is the wavelength of the wave?

A standing wave on a guitar string with two nodes and three antinodes means that the string is divided into three equal segments.

Each segment has one antinode and one node.

The wavelength of the wave is twice the length of one segment.

The wavelength is λ = 2 x (60 cm / 3) = 40 cm.

Answer: the wavelength of the wave on the guitar string is 40 cm.

Exercise 3

Questions

a) A water wave has a frequency of 0.5 Hz and a wavelength of 2 m. What is the wave speed?

b) A light wave has a frequency of 6 x 10^14 Hz and a wavelength of 500 nm. What is the wavelength of a photon of this light?

c) A sound wave travels through air with a speed of 340 m/s. What is the wavelength of the wave if its frequency is 100 Hz?

d) A light wave has a wavelength of 600 nm. What is the frequency of the wave?

e) A guitar string is plucked and produces a frequency of 440 Hz. If the length of the string is 80 cm, what is the wavelength of the wave?

Answers

a) A water wave has a frequency of 0.5 Hz and a wavelength of 2 m. What is the wave speed?

v = fλ, where v is the speed of the wave, f is the frequency, and λ is the wavelength.

v = 0.5 x 2 = 1 m/s.

Answer: the wave speed of the water wave is 1 m/s.

b) A light wave has a frequency of 6 x 10^14 Hz and a wavelength of 500 nm. What is the wavelength of a photon of this light?

$E = hf = hc/\lambda$, where E is the energy of the photon, h is Planck's constant, c is the speed of light, and λ is the wavelength of the photon.

$\lambda = hc/E = (6.626 \times 10^{-34} \text{ J.s})(3 \times 10^8 \text{ m/s})/(6 \times 10^{14} \text{ Hz}) = 3.31 \times 10^{-19}$ m.

Answer: the wavelength of a photon of this light is 3.31 x 10^-19 m.

c) A sound wave travels through air with a speed of 340 m/s. What is the wavelength of the wave if its frequency is 100 Hz?

$v = f\lambda$, where v is the speed of the wave, f is the frequency, and λ is the wavelength.

$\lambda = v/f = 340/100 = 3.4$ m.

Answer: the wavelength of the sound wave is 3.4 meters.

d) A light wave has a wavelength of 600 nm. What is the frequency of the wave?

$v = f\lambda$, where v is the speed of light (3 x 10^8 m/s), λ is the wavelength, and f is the frequency.

$f = v/\lambda = (3 \times 10^8 \text{ m/s})/(600 \times 10^{-9} \text{ m}) = 5 \times 10^{14}$ Hz.

Answer: the frequency of the light wave is 5 x 10^14 Hz.

e) A guitar string is plucked and produces a frequency of 440 Hz. If the length of the string is 80 cm, what is the wavelength of the wave?

$v = f\lambda$, where v is the speed of the wave, f is the frequency, and λ is the wavelength.

To find the wavelength, we need to first find the speed of the wave.

The speed of the wave depends on the tension in the string and the mass per unit length of the string.

Since these values are not given, we will assume a typical value for the speed of sound in steel strings, which is about 5000 m/s.

$\lambda = v/f = 5000$ m

Exercise 4

Questions

a) A wave traveling on a string has a frequency of 20 Hz and an amplitude of 5 cm. What is the maximum speed of the string during one cycle of the wave?

b) A sound wave with a frequency of 500 Hz travels through air with a speed of 340 m/s. What is the wavelength of the wave?

c) A water wave has a wavelength of 5 meters and a period of 10 seconds. What is the frequency of the wave?

d) A light wave travels through a vacuum with a speed of 3 x 10^8 m/s. If the wavelength of the wave is 500 nm, what is the frequency of the wave?

e) A wave on a string has a wavelength of 2 meters and a speed of 20 m/s. What is the frequency of the wave?

Answers

a) A wave traveling on a string has a frequency of 20 Hz and an amplitude of 5 cm. What is the maximum speed of the string during one cycle of the wave?

vmax = Aω, where A is the amplitude of the oscillation and ω is the angular frequency.

The angular frequency is related to the frequency by the equation ω = 2πf.

vmax = (5 cm)(2π)(20 Hz) = 628 cm/s.

Answer: the maximum speed of the string during one cycle of the wave is 628 cm/s.

b) A sound wave with a frequency of 500 Hz travels through air with a speed of 340 m/s. What is the wavelength of the wave?

v = fλ, where v is the speed of the wave, f is the frequency, and λ is the wavelength.

λ = v/f = 340/500 = 0.68 m.

Answer: the wavelength of the sound wave is 0.68 meters.

c) A water wave has a wavelength of 5 meters and a period of 10 seconds. What is the frequency of the wave?

T = 1/f, where T is the period and f is the frequency.

f = 1/T = 1/10 = 0.1 Hz.

Answer: the frequency of the water wave is 0.1 Hz.

d) A light wave travels through a vacuum with a speed of 3 x 10^8 m/s. If the wavelength of the wave is 500 nm, what is the frequency of the wave?

$v = f\lambda$, where v is the speed of light, λ is the wavelength, and f is the frequency.

$f = v/\lambda = (3 \times 10^8$ m/s$)/(500 \times 10^{-9}$ m$) = 6 \times 10^{14}$ Hz.

Answer: the frequency of the light wave is 6×10^{14} Hz.

e) A wave on a string has a wavelength of 2 meters and a speed of 20 m/s. What is the frequency of the wave?

$v = f\lambda$, where v is the speed of the wave, λ is the wavelength, and f is the frequency.

$f = v/\lambda = 20/2 = 10$ Hz.

Answer: the frequency of the wave on the string is 10 Hz.

Exercise 5

Questions

a) A sound wave with a frequency of 1000 Hz and a wavelength of 34 cm travels through air. What is the speed of the wave?

b) A wave on a string has a wavelength of 0.5 meters and a frequency of 100 Hz. What is the speed of the wave?

c) A wave on a string has a frequency of 60 Hz and a wavelength of 1 meter. What is the period of the wave?

d) An electromagnetic wave has a wavelength of 500 nm and a frequency of 6 x 10^14 Hz. What is the speed of light?

e) A wave on a string has an amplitude of 8 cm and a wavelength of 0.5 meters. What is the maximum displacement of a particle on the string?

Answers

a) A sound wave with a frequency of 1000 Hz and a wavelength of 34 cm travels through air. What is the speed of the wave?

$v = f\lambda$, where v is the speed of the wave, f is the frequency, and λ is the wavelength.

$v = f\lambda = (1000$ Hz$)(34$ cm$) = 34000$ cm/s $= 340$ m/s.

Answer: the speed of the sound wave is 340 m/s.

b) A wave on a string has a wavelength of 0.5 meters and a frequency of 100 Hz. What is the speed of the wave?

$v = f\lambda$, where v is the speed of the wave, f is the frequency, and λ is the wavelength.

$v = f\lambda = (100 \text{ Hz})(0.5 \text{ meters}) = 50 \text{ m/s}$

Answer: the speed of the wave is 50 m/s

c) A wave on a string has a frequency of 60 Hz and a wavelength of 1 meter. What is the period of the wave?

$T = 1/f$, where T is the period and f is the frequency.

$T = 1/60 = 0.0167$ seconds.

Answer: the period of the wave is 0.0167 seconds.

d) An electromagnetic wave has a wavelength of 500 nm and a frequency of 6 x 10^14 Hz. What is the speed of light?

$v = f\lambda$, where v is the speed of light, f is the frequency, and λ is the wavelength.

$v = f\lambda = (6 \times 10^{14} \text{ Hz})(500 \text{ nm}) = 3 \times 10^8 \text{ m/s}$.

Answer: the speed of light is 3 x 10^8 m/s.

e) A wave on a string has an amplitude of 8 cm and a wavelength of 0.5 meters. What is the maximum displacement of a particle on the string?

We can use the equation for the displacement of a particle on a string in terms of the amplitude and position, which is :

$y(x,t) = A*\sin(kx - \omega t)$, where k is the wave number and ω is the angular frequency.

The maximum displacement occurs when $\sin(kx - \omega t) = 1$, which means $y(x,t) = A$. Substituting the values, we get $y(x,t) = 8$ cm.

Answer: the maximum displacement of a particle on the string is 8 cm.

Part 2: Waves

Exercise 1

Questions

a) A sound wave with a frequency of 400 Hz and an intensity of 50 W/m^2 travels through air. What is the amplitude of the wave?

b) A wave on a string has a wavelength of 0.8 meters and a speed of 16 m/s. What is the wave number of the wave?

c) A wave on a string has a wavelength of 0.6 meters and a frequency of 50 Hz. What is the speed of the wave?

d) An electromagnetic wave has a wavelength of 1.5 μm and a frequency of 2 x 10^14 Hz. What is the energy of one photon of the wave?

e) A sound wave with a frequency of 200 Hz and an amplitude of 2 mm travels through air. What is the intensity of the wave?

Answers

a) A sound wave with a frequency of 400 Hz and an intensity of 50 W/m^2 travels through air. What is the amplitude of the wave?

I = (1/2)ρvω^2A^2, where ρ is the density of the medium, v is the speed of sound, ω is the angular frequency, and A is the amplitude.

A = sqrt(2I/(ρvω^2)) = sqrt((2)(50 W/m^2)/((1.2 kg/m^3)(343 m/s)(2π(400 Hz))^2)) = 2.75 x 10^-5 m.

Answer: the amplitude of the sound wave is 2.75 x 10^-5 m.

b) A wave on a string has a wavelength of 0.8 meters and a speed of 16 m/s. What is the wave number of the wave?

k = 2π/λ, where k is the wave number and λ is the wavelength.

k = 2π/0.8 = 7.85 m^-1.

Answer: the wave number of the wave on the string is 7.85 m^-1.

c) A wave on a string has a wavelength of 0.6 meters and a frequency of 50 Hz. What is the speed of the wave?

v = fλ, where v is the speed of the wave, f is the frequency, and λ is the wavelength.

v = fλ = (50 Hz)(0.6 meters) = 30 m/s.

Answer: the speed of the wave on the string is 30 m/s.

d) An electromagnetic wave has a wavelength of 1.5 μm and a frequency of 2 x 10^14 Hz. What is the energy of one photon of the wave?

We can use the formula for the energy of a photon in terms of its frequency, which is E = hf, where h is Planck's constant.

We can also use the formula for the frequency of a wave in terms of its wavelength, which is f = c/λ, where c is the speed of light.

E = hc/λ = (6.626 x 10^-34 J s)(2 x 10^14 Hz) / (1.5 x 10^-6 m) = 8.84 x 10^-19 J.

Answer: the energy of one photon of the electromagnetic wave is 8.84 x 10^-19 J.

e) A sound wave with a frequency of 200 Hz and an amplitude of 2 mm travels through air. What is the intensity of the wave?

We can use the formula for the intensity of a sound wave in terms of the amplitude and frequency, which is I = (1/2)ρvω^2A^2, where ρ is the density of the medium, v is the speed of sound, ω is the angular frequency, and A is the amplitude.

I = (1/2)(1.2 kg/m^3)(343 m/s)(2π(200 Hz))^2(2 x 10^-3 m)^2 = 0.19 W/m^2.

Answer: the intensity of the sound wave is 0.19 W/m^2.

Exercise 2

Questions

a) A wave on a string has an amplitude of 10 cm and a frequency of 80 Hz. What is the maximum speed of a particle on the string?

b) An electromagnetic wave has a wavelength of 500 nm and a speed of 3 x 10^8 m/s. What is the frequency of the wave?

c) A sound wave with a wavelength of 0.5 meters and a frequency of 1000 Hz travels through air. What is the phase difference between two points on the wave that are separated by 0.2 meters?

d) A wave on a string has a frequency of 60 Hz and a wavelength of 0.5 meters. What is the period of the wave?

e) An electromagnetic wave has an electric field amplitude of 100 V/m and a magnetic field amplitude of 0.1 T. What is the intensity of the wave?

Answers

a) A wave on a string has an amplitude of 10 cm and a frequency of 80 Hz. What is the maximum speed of a particle on the string?

We can use the equation for the velocity of a particle on a string in terms of the amplitude and position, which is $v(x,t) = \omega A^*\cos(kx - \omega t)$, where k is the wave number and ω is the angular frequency.

The maximum speed occurs when $\cos(kx - \omega t) = 1$, which means $v(x,t) = \omega A$.

$v(x,t) = (2\pi(80\ Hz))(10\ cm) = 502\ cm/s$.

Answer: the maximum speed of a particle on the string is 502 cm/s.

b) An electromagnetic wave has a wavelength of 500 nm and a speed of 3 x 10^8 m/s. What is the frequency of the wave?

$f = v/\lambda$, where f is the frequency, v is the speed of the wave, and λ is the wavelength.

$f = (3 \times 10^8\ m/s)/(500\ nm) = 6 \times 10^{14}\ Hz$.

Answer: the frequency of the electromagnetic wave is 6 x 10^14 Hz.

c) A sound wave with a wavelength of 0.5 meters and a frequency of 1000 Hz travels through air. What is the phase difference between two points on the wave that are separated by 0.2 meters?

We can use the formula for the phase difference between two points on a wave in terms of the wave number and the distance between the points, which is $\Delta\varphi = k\Delta x$, where k is the wave number and Δx is the distance between the points.

$k = 2\pi/\lambda$, where λ is the wavelength.

$k = 2\pi/(0.5\ meters) = 12.56\ m^{-1}$, and $\Delta\varphi = (12.56\ m^{-1})(0.2\ meters) = 2.51$ radians.

Answer: the phase difference between two points on the sound wave that are separated by 0.2 meters is 2.51 radians.

d) A wave on a string has a frequency of 60 Hz and a wavelength of 0.5 meters. What is the period of the wave?

T = 1/f

T = 1/60 Hz = 0.0167 seconds.

Answer: the period of the wave on the string is 0.0167 seconds.

e) An electromagnetic wave has an electric field amplitude of 100 V/m and a magnetic field amplitude of 0.1 T. What is the intensity of the wave?

We can use the formula for the intensity of an electromagnetic wave in terms of the electric and magnetic field amplitudes, which is I = (1/2)εcE^2 = (1/2)cB^2/μ, where ε is the permittivity of free space, c is the speed of light, μ is the permeability of free space, E is the electric field amplitude, and B is the magnetic field amplitude.

I = (1/2)(8.85 x 10^-12 F/m)(3 x 10^8 m/s)(100 V/m)^2 = 1.33 x 10^-6 W/m^2.

Answer: the intensity of the electromagnetic wave is 1.33 x 10^-6 W/m^2.

Exercise 3

Questions

a) A sound wave with a frequency of 500 Hz and an amplitude of 5 cm travels through air. What is the wavelength of the wave?

b) A wave on a string has a velocity of 20 m/s and a wavelength of 0.4 meters. What is the frequency of the wave?

c) An electromagnetic wave has a frequency of 3 x 10^16 Hz and a wavelength of 10 nm. What is the energy of one photon of the wave?

d) A wave has a velocity of 340 m/s and a frequency of 500 Hz. What is the wavelength of the wave?

e) A sound wave with a wavelength of 0.3 meters and an amplitude of 10 cm travels through air. What is the maximum displacement of a particle in the air due to the wave?

Answers

a) A sound wave with a frequency of 500 Hz and an amplitude of 5 cm travels through air. What is the wavelength of the wave?

$\lambda = v/f$, where λ is the wavelength, v is the speed of sound, and f is the frequency.

$\lambda = (343 \text{ m/s})/(500 \text{ Hz}) = 0.686$ meters.

Answer: the wavelength of the sound wave is 0.686 meters.

b) A wave on a string has a velocity of 20 m/s and a wavelength of 0.4 meters. What is the frequency of the wave?

$f = v/\lambda$, where f is the frequency, v is the velocity of the wave, and λ is the wavelength.

$f = (20 \text{ m/s})/(0.4 \text{ meters}) = 50$ Hz.

Answer: the frequency of the wave on the string is 50 Hz.

c) An electromagnetic wave has a frequency of 3×10^{16} Hz and a wavelength of 10 nm. What is the energy of one photon of the wave?

$E = hf$, where E is the energy of the photon, h is Planck's constant, and f is the frequency.

$\lambda = c/f$, where λ is the wavelength and c is the speed of light.

$f = c/\lambda$.

$E = hc/\lambda$.

$E = (6.626 \times 10^{-34} \text{ J s})(3 \times 10^{8} \text{ m/s})/(10 \times 10^{-9} \text{ m}) = 1.988 \times 10^{-15}$ J.

Answer: the energy of one photon of the electromagnetic wave is 1.988×10^{-15} J.

d) A wave has a velocity of 340 m/s and a frequency of 500 Hz. What is the wavelength of the wave?

$\lambda = v/f$, where λ is the wavelength, v is the velocity of the wave, and f is the frequency.

$\lambda = (340 \text{ m/s})/(500 \text{ Hz}) = 0.68$ meters.

Answer: the wavelength of the wave is 0.68 meters.

e) A sound wave with a wavelength of 0.3 meters and an amplitude of 10 cm travels through air. What is the maximum displacement of a particle in the air due to the wave?

We can use the formula for the maximum displacement of a particle in a sound wave in terms of its amplitude, which is $\Delta x = A$.

$\Delta x = (10 \text{ cm}) = 0.1$ meters.

Answer: the maximum displacement of a particle in the air due to the sound wave is 0.1 meters.

Exercise 4

Questions

a) A wave on a string has an amplitude of 5 mm and a wavelength of 0.2 meters. What is the maximum speed of a particle on the string due to the wave?

b) An electromagnetic wave has a wavelength of 500 nanometers and a frequency of 6 x 10^14 Hz. What is the speed of the wave?

c) A wave on a string has a frequency of 50 Hz and a wavelength of 0.5 meters. What is the period of the wave?

d) A sound wave has a frequency of 440 Hz and a wavelength of 0.8 meters. What is the period of the wave?

e) A wave on a string has an amplitude of 2 cm and a wavelength of 0.4 meters. What is the maximum acceleration of a particle on the string due to the wave?

Answers

a) A wave on a string has an amplitude of 5 mm and a wavelength of 0.2 meters. What is the maximum speed of a particle on the string due to the wave?

vmax = 2πAf, where vmax is the maximum speed of the particle, A is the amplitude of the wave, and f is the frequency.

f = v/λ, where v is the velocity of the wave and λ is the wavelength.

v = fλ

vmax = 2πA(v/λ)

vmax = 2π(5 x 10^-3 m)(v/(0.2 meters)) = (31.4 m/s)v.

vmax = (1/31.4 m/s)(5 x 10^-3 m)(v) = 0.000159 m/s (v).

Answer: the maximum speed of a particle on the string due to the wave is 0.000159 m/s times the velocity of the wave.

b) An electromagnetic wave has a wavelength of 500 nanometers and a frequency of 6 x 10^14 Hz. What is the speed of the wave?

v = λf, where v is the speed of the wave, λ is the wavelength, and f is the frequency.

v = (500 x 10^-9 meters)(6 x 10^14 Hz) = 3 x 10^8 m/s.

Answer: the speed of the electromagnetic wave is 3 x 10^8 m/s.

c) A wave on a string has a frequency of 50 Hz and a wavelength of 0.5 meters. What is the period of the wave?

$T = 1/f$, where T is the period and f is the frequency.

$T = 1/(50 \text{ Hz}) = 0.02$ seconds.

Answer: the period of the wave is 0.02 seconds.

d) A sound wave has a frequency of 440 Hz and a wavelength of 0.8 meters. What is the period of the wave?

$T = 1/f$, where T is the period and f is the frequency.

$T = 1/(440 \text{ Hz}) = 0.00227$ seconds.

Answer: the period of the wave is 0.00227 seconds.

e) A wave on a string has an amplitude of 2 cm and a wavelength of 0.4 meters. What is the maximum acceleration of a particle on the string due to the wave?

$a_{max} = 4\pi^2 A f^2$, where amax is the maximum acceleration of the particle, A is the amplitude of the wave, and f is the frequency.

$f = v/\lambda$, where v is the velocity of the wave and λ is the wavelength.

$v = f\lambda$.

$a_{max} = 4\pi^2 A(v/\lambda)^2$.

$a_{max} = 4\pi^2(2 \times 10^{-2} \text{ m})(v/(0.4 \text{ meters}))^2$.

$a_{max} = (1/(4\pi^2(2 \times 10^{-2} \text{ m})))v^2$.

Answer: the maximum acceleration of a particle on the string due to the wave is $(1/(4\pi^2(2 \times 10^{-2} \text{ m})))$ times the square of the velocity of the wave.

Exercise 5

Questions

a) A wave has a wavelength of 1.2 meters and a frequency of 300 Hz. What is the period of the wave?

b) A wave has a wavelength of 0.6 meters and a velocity of 10 m/s. What is the frequency of the wave?

c) A wave on a string has a wavelength of 1 meter and a frequency of 100 Hz. What is the speed of the wave?

d) A wave has a frequency of 20 Hz and a speed of 5 m/s. What is the wavelength of the wave?

e) A sound wave has a wavelength of 1.5 meters and a velocity of 340 m/s. What is the frequency of the wave?

Answers

a) A wave has a wavelength of 1.2 meters and a frequency of 300 Hz. What is the period of the wave?

T = 1/f, where T is the period and f is the frequency. Substituting the value, we get T = 1/(300 Hz) = 0.00333 seconds.

Answer: the period of the wave is 0

b) A wave has a wavelength of 0.6 meters and a velocity of 10 m/s. What is the frequency of the wave?

f = v/λ, where f is the frequency, v is the velocity of the wave, and λ is the wavelength.

f = 10 m/s / 0.6 meters = 16.67 Hz.

Answer: the frequency of the wave is 16.67 Hz.

c) A wave on a string has a wavelength of 1 meter and a frequency of 100 Hz. What is the speed of the wave?

v = sqrt(T/μ), where v is the speed of the wave, T is the tension in the string, and μ is the linear density of the string.

f = v/λ, where f is the frequency of the wave, λ is its wavelength, and v is its velocity.

v = λf

v = 1 meter x 100 Hz = 100 m/s.

Answer: the speed of the wave is 100 m/s.

d) A wave has a frequency of 20 Hz and a speed of 5 m/s. What is the wavelength of the wave?

λ = v/f, where λ is the wavelength, v is the velocity of the wave, and f is its frequency.

λ = 5 m/s / 20 Hz = 0.25 meters.

Answer: the wavelength of the wave is 0.25 meters.

e) A sound wave has a wavelength of 1.5 meters and a velocity of 340 m/s. What is the frequency of the wave?

$f = v/\lambda$, where f is the frequency of the wave, v is its velocity, and λ is its wavelength.

f = 340 m/s / 1.5 meters = 226.67 Hz.

Answer: the frequency of the sound wave is 226.67 Hz.

Part 3: Waves

Exercise 1

Questions

a) A wave on a string has a frequency of 60 Hz and a speed of 120 m/s. What is the wavelength of the wave?

b) A wave on a string has a wavelength of 0.5 meters and a frequency of 200 Hz. What is the period of the wave?

c) A sound wave has a wavelength of 2 meters and a frequency of 250 Hz. What is the speed of the wave?

d) A wave on a string has a wavelength of 1 meter and a speed of 40 m/s. What is the frequency of the wave?

Answers

a) A wave on a string has a frequency of 60 Hz and a speed of 120 m/s. What is the wavelength of the wave?

$\lambda = v/f$, where λ is the wavelength, v is the velocity of the wave, and f is its frequency.

λ = 120 m/s / 60 Hz = 2 meters.

Answer: the wavelength of the wave is 2 meters.

b) A wave on a string has a wavelength of 0.5 meters and a frequency of 200 Hz. What is the period of the wave?

$T = 1/f$, where T is the period of the wave and f is its frequency.

T = 1/200 Hz = 0.005 seconds.

Answer: the period of the wave is 0.005 seconds.

c) A sound wave has a wavelength of 2 meters and a frequency of 250 Hz. What is the speed of the wave?

$v = \lambda f$, where v is the speed of the wave, λ is its wavelength, and f is its frequency.

v = 2 meters x 250 Hz = 500 m/s.

Answer: the speed of the sound wave is 500 m/s.

d) A wave on a string has a wavelength of 1 meter and a speed of 40 m/s. What is the frequency of the wave?

$f = v/\lambda$, where f is the frequency of the wave, v is the velocity of the wave, and λ is its wavelength.

f = 40 m/s / 1 meter = 40 Hz.

Answer: the frequency of the wave is 40 Hz.

Exercise 2

Questions

a) A wave has a frequency of 50 Hz and a wavelength of 0.8 meters. What is the period of the wave?

b) A wave on a string has a frequency of 100 Hz and a wavelength of 0.5 meters. What is the speed of the wave?

c) A wave has a wavelength of 0.2 meters and a speed of 300 m/s. What is the frequency of the wave?

d) A guitar string has a length of 0.75 meters and is under a tension of 200 N. If the mass per unit length of the string is 0.02 kg/m, what is the speed of a wave on the string?

e) A sound wave has a frequency of 1000 Hz and a velocity of 340 m/s. What is the wavelength of the wave?

Answers

a) A wave has a frequency of 50 Hz and a wavelength of 0.8 meters. What is the period of the wave?

T = 1/f, where T is the period of the wave and f is its frequency.

T = 1/50 Hz = 0.02 seconds.

Answer: the period of the wave is 0.02 seconds.

b) A wave on a string has a frequency of 100 Hz and a wavelength of 0.5 meters. What is the speed of the wave?

$v = sqrt(T/\mu)$, where v is the speed of the wave, T is the tension in the string, and μ is the linear density of the string.

$f = v/\lambda$, where f is the frequency of the wave, λ is its wavelength, and v is its velocity.

$v = \lambda f$

v = 0.5 meters x 100 Hz = 50 m/s.

Answer: the speed of the wave is 50 m/s.

c) A wave has a wavelength of 0.2 meters and a speed of 300 m/s. What is the frequency of the wave?

$f = v/\lambda$, where f is the frequency of the wave, v is its velocity, and λ is its wavelength.

f = 300 m/s / 0.2 meters = 1500 Hz.

Answer: the frequency of the wave is 1500 Hz.

d) A guitar string has a length of 0.75 meters and is under a tension of 200 N. If the mass per unit length of the string is 0.02 kg/m, what is the speed of a wave on the string?

$v = \sqrt{T/\mu}$, where v is the speed of the wave, T is the tension in the string, and μ is the linear density of the string.

$v = \sqrt{200\ N / 0.02\ kg/m} = 20$ m/s.

Answer: the speed of a wave on the string is 20 m/s.

e) A sound wave has a frequency of 1000 Hz and a velocity of 340 m/s. What is the wavelength of the wave?

$\lambda = v/f$, where λ is the wavelength of the wave, v is its velocity, and f is its frequency.

λ = 340 m/s / 1000 Hz = 0.34 meters.

Answer: the wavelength of the sound wave is 0.34 meters.

Exercise 3

Questions

a) Two waves with the same amplitude and wavelength are traveling in opposite directions on a string. If the amplitude of each wave is 2 cm, what is the amplitude of the resulting wave?

b) A wave has an amplitude of 3 cm and a wavelength of 0.6 meters. What is the maximum displacement of a particle in the medium from its equilibrium position?

c) A wave on a string has a wavelength of 0.4 meters and a speed of 20 m/s. What is the frequency of the wave?

d) A transverse wave is traveling along a rope with a frequency of 50 Hz and a wavelength of 0.4 meters. If the wave has an amplitude of 8 cm, what is the maximum velocity of a particle in the rope?

Answers

a) Two waves with the same amplitude and wavelength are traveling in opposite directions on a string. If the amplitude of each wave is 2 cm, what is the amplitude of the resulting wave?

When two waves of the same amplitude and wavelength travel in opposite directions, they interfere destructively and create a standing wave with zero amplitude at the nodes and twice the amplitude of each wave at the antinodes. The amplitude of the resulting wave is 2 x 2 cm = 4 cm.

Answer: the amplitude of the resulting wave is 4 cm.

b) A wave has an amplitude of 3 cm and a wavelength of 0.6 meters. What is the maximum displacement of a particle in the medium from its equilibrium position?

The maximum displacement of a particle in a medium from its equilibrium position is equal to the amplitude of the wave.

The maximum displacement is 3 cm.

Answer: the maximum displacement of a particle in the medium from its equilibrium position is 3 cm.

c) A wave on a string has a wavelength of 0.4 meters and a speed of 20 m/s. What is the frequency of the wave?

$f = v/\lambda$, where f is the frequency of the wave, v is its velocity, and λ is its wavelength.

f = 20 m/s / 0.4 meters = 50 Hz.

Answer: the frequency of the wave is 50 Hz.

d) A transverse wave is traveling along a rope with a frequency of 50 Hz and a wavelength of 0.4 meters. If the wave has an amplitude of 8 cm, what is the maximum velocity of a particle in the rope?

v_max = f * λ * A, where v_max is the maximum velocity, f is the frequency, λ is the wavelength, and A is the amplitude.

v_max = 50 Hz * 0.4 m * 8 cm = 16 m/s.

Answer: the maximum velocity of a particle in the rope is 16 m/s.

Exercise 4

Questions

a) A longitudinal sound wave has a frequency of 440 Hz and a wavelength of 0.75 meters. What is the speed of sound in the medium?

b) Two waves with the same amplitude and frequency are traveling in the same direction on a string. If the amplitude of each wave is 2 mm, what is the amplitude of the resulting wave?

c) A wave on a string has a frequency of 100 Hz and a speed of 20 m/s. What is the wavelength of the wave?

d) A wave on a string has a frequency of 50 Hz and a wavelength of 0.8 meters. What is the speed of the wave?

e) A transverse wave is traveling along a rope with a frequency of 60 Hz and a wavelength of 0.6 meters. If the wave has an amplitude of 6 mm, what is the maximum displacement of a particle in the rope?

Answers

a) A longitudinal sound wave has a frequency of 440 Hz and a wavelength of 0.75 meters. What is the speed of sound in the medium?

v = f * λ, where v is the speed of sound, f is the frequency of the wave, and λ is its wavelength.

v = 440 Hz * 0.75 m = 330 m/s.

Answer: the speed of sound in the medium is 330 m/s.

b) Two waves with the same amplitude and frequency are traveling in the same direction on a string. If the amplitude of each wave is 2 mm, what is the amplitude of the resulting wave?

When two waves of the same amplitude and frequency travel in the same direction, they interfere constructively and create a wave with twice the amplitude of each wave.

The amplitude of the resulting wave is 2 * 2 mm = 4 mm.

Answer: the amplitude of the resulting wave is 4 mm.

c) A wave on a string has a frequency of 100 Hz and a speed of 20 m/s. What is the wavelength of the wave?

λ = v/f, where λ is the wavelength of the wave, v is its speed, and f is its frequency.

λ = 20 m/s / 100 Hz = 0.2 m

Answer: the wavelength of the wave is 0.2 m.

d) A wave on a string has a frequency of 50 Hz and a wavelength of 0.8 meters. What is the speed of the wave?

v = f * λ, where v is the speed of the wave, f is its frequency, and λ is its wavelength.

v = 50 Hz * 0.8 m = 40 m/s.

Answer: the speed of the wave is 40 m/s.

e) A transverse wave is traveling along a rope with a frequency of 60 Hz and a wavelength of 0.6 meters. If the wave has an amplitude of 6 mm, what is the maximum displacement of a particle in the rope?

The maximum displacement of a particle in a rope is equal to the amplitude of the wave.

The maximum displacement is 6 mm.

Answer: the maximum displacement of a particle in the rope is 6 mm.

Exercise 5

Questions

a) A wave with a frequency of 200 Hz and a wavelength of 2 meters is traveling in a medium. What is the period of the wave?

b) Two waves of the same frequency and amplitude travel in opposite directions on a string. What is the amplitude of the resulting wave at a point where the two waves interfere destructively?

c) A water wave has a frequency of 10 Hz and a speed of 5 m/s. What is the wavelength of the wave?

d) A wave on a string has a frequency of 80 Hz and a wavelength of 0.5 meters. What is the period of the wave?

e) A wave with a frequency of 1000 Hz and a wavelength of 0.1 meters is traveling in a medium. What is the speed of the wave?

Answer

a) A wave with a frequency of 200 Hz and a wavelength of 2 meters is traveling in a medium. What is the period of the wave?

$T = 1/f$, where T is the period of the wave and f is its frequency. Substituting the given values, we get $T = 1/200$ Hz = 0.005 seconds.

Answer: the period of the wave is 0.005 seconds.

b) Two waves of the same frequency and amplitude travel in opposite directions on a string. What is the amplitude of the resulting wave at a point where the two waves interfere destructively?

When two waves of the same frequency and amplitude travel in opposite directions, they interfere destructively and cancel each other out at certain points.

At these points, the resulting wave has zero amplitude.

Therefore, the amplitude of the resulting wave at a point where the two waves interfere destructively is zero.

c) A water wave has a frequency of 10 Hz and a speed of 5 m/s. What is the wavelength of the wave?

$\lambda = v/f$, where λ is the wavelength of the wave, v is its speed, and f is its frequency.

$\lambda = 5$ m/s / 10 Hz = 0.5 m.

Answer: the wavelength of the wave is 0.5 m.

d) A wave on a string has a frequency of 80 Hz and a wavelength of 0.5 meters. What is the period of the wave?

T = 1/f, where T is the period of the wave and f is its frequency.

T = 1/80 Hz = 0.0125 seconds.

Answer: the period of the wave is 0.0125 seconds.

e) A wave with a frequency of 1000 Hz and a wavelength of 0.1 meters is traveling in a medium. What is the speed of the wave?

v = f * λ, where v is the speed of the wave, f is its frequency, and λ is its wavelength.

v = 1000 Hz * 0.1 m = 100 m/s.

Answer: the speed of the wave is 100 m/s.

Part 4: Waves

Exercise 1

Questions

a) A transverse wave is traveling along a rope with a frequency of 40 Hz and a wavelength of 0.8 meters. If the wave has an amplitude of 5 cm, what is the maximum displacement of a particle in the rope?

b) A sound wave has a frequency of 500 Hz and a wavelength of 0.7 meters. What is the speed of the sound wave?

c) A wave has a frequency of 120 Hz and a speed of 360 m/s. What is the wavelength of the wave?

d) A wave is traveling in a medium with a speed of 200 m/s. If the wavelength of the wave is 0.5 meters and its amplitude is 0.1 meters, what is the maximum velocity of a particle in the medium?

e) A wave is traveling in a medium with a speed of 300 m/s. If the wavelength of the wave is 2 meters and its amplitude is 0.2 meters, what is the maximum acceleration of a particle in the medium?

Answers

a) A transverse wave is traveling along a rope with a frequency of 40 Hz and a wavelength of 0.8 meters. If the wave has an amplitude of 5 cm, what is the maximum displacement of a particle in the rope?

The maximum displacement of a particle in a rope is equal to the amplitude of the wave.

Therefore, we can convert the amplitude to meters and use the formula max displacement = amplitude * 2, where max displacement is the maximum displacement of a particle in the rope.

Converting the amplitude to meters, we get 0.05 m.

max displacement = 0.05 m * 2 = 0.1 m.

Answer: the maximum displacement of a particle in the rope is 0.1 m.

b) A sound wave has a frequency of 500 Hz and a wavelength of 0.7 meters. What is the speed of the sound wave?

$v = f * \lambda$, where v is the speed of the sound wave, f is its frequency, and λ is its wavelength.

v = 500 Hz * 0.7 m = 350 m/s.

Answer: the speed of the sound wave is 350 m/s.

c) A wave has a frequency of 120 Hz and a speed of 360 m/s. What is the wavelength of the wave?

$\lambda = v/f$, where λ is the wavelength of the wave, v is its speed, and f is its frequency.

λ = 360 m/s / 120 Hz = 3 meters.

Answer: the wavelength of the wave is 3 meters.

d) A wave is traveling in a medium with a speed of 200 m/s. If the wavelength of the wave is 0.5 meters and its amplitude is 0.1 meters, what is the maximum velocity of a particle in the medium?

The maximum velocity of a particle in a medium is equal to the product of the amplitude of the wave and its angular frequency, which is equal to 2π times its frequency.

max velocity = amplitude * angular frequency, where max velocity is the maximum velocity of a particle in the medium. Converting the amplitude to meters, we get 0.1 m.

$v = f * \lambda$, where v is the speed of the wave, λ is its wavelength, and f is its frequency.

$f = v/\lambda$ = 200 m/s / 0.5 m = 400 Hz.

The angular frequency is then 2π times the frequency, which is 2π * 400 Hz = 2513.27 rad/s.

max velocity = 0.1 m * 2513.27 rad/s = 251.327 m/s.

Answer: the maximum velocity of a particle in the medium is 251.327 m/s.

e) A wave is traveling in a medium with a speed of 300 m/s. If the wavelength of the wave is 2 meters and its amplitude is 0.2 meters, what is the maximum acceleration of a particle in the medium?

The maximum acceleration of a particle in a medium is equal to the product of the amplitude of the wave and its angular frequency squared.

max acceleration = amplitude * angular frequency^2, where max acceleration is the maximum acceleration of a particle in the medium.

Converting the amplitude to meters, we get 0.2 m.

v = f * λ, where v is the speed of the wave, λ is its wavelength, and f is its frequency.

f = v/λ = 300 m/s / 2 m = 150 Hz.

The angular frequency is then 2π times the frequency, which is 2π * 150 Hz = 942.48 rad/s.

max acceleration = 0.2 m * (942.48 rad/s)^2 = 176814.31 m/s^2.

Answer: the maximum acceleration of a particle in the medium is 176814.31 m/s^2.

Exercise 2

Questions

a) A wave with a wavelength of 0.1 meters and a frequency of 1000 Hz is traveling in a medium with a speed of 100 m/s. What is the period of the wave?

b) A wave is traveling in a medium with a frequency of 200 Hz and a wavelength of 2 meters. If the maximum displacement of the wave is 0.1 meters, what is the amplitude of the wave?

c) A wave is traveling in a medium with a speed of 400 m/s and a wavelength of 0.1 meters. What is the frequency of the wave?

d) A wave is traveling in a medium with a wavelength of 0.5 meters and an amplitude of 0.2 meters. What is the maximum energy density of the wave?

e) A wave is traveling in a medium with a frequency of 300 Hz and a speed of 600 m/s. What is the wavelength of the wave?

Answers

a) A wave with a wavelength of 0.1 meters and a frequency of 1000 Hz is traveling in a medium with a speed of 100 m/s. What is the period of the wave?

T = 1/f, where T is the period of the wave and f is its frequency.

T = 1/1000 Hz = 0.001 seconds.

Answer: the period of the wave is 0.001 seconds.

b) A wave is traveling in a medium with a frequency of 200 Hz and a wavelength of 2 meters. If the maximum displacement of the wave is 0.1 meters, what is the amplitude of the wave?

The amplitude of a wave is half of its maximum displacement.

The formula for the amplitude of a wave in terms of its maximum displacement, is A = 0.5 * max displacement, where A is the amplitude of the wave and max displacement is its maximum displacement.

A = 0.5 * 0.1 meters = 0.05 meters.

Answer: the amplitude of the wave is 0.05 meters.

c) A wave is traveling in a medium with a speed of 400 m/s and a wavelength of 0.1 meters. What is the frequency of the wave?

$f = v/\lambda$, where f is the frequency of the wave, v is its speed, and λ is its wavelength.

f = 400 m/s / 0.1 m = 4000 Hz.

Answer: the frequency of the wave is 4000 Hz.

d) A wave is traveling in a medium with a wavelength of 0.5 meters and an amplitude of 0.2 meters. What is the maximum energy density of the wave?

The energy density of a wave is proportional to the square of its amplitude.

The formula for the maximum energy density of a wave is $(1/2) * \rho * (A^2)$, where ρ is the density of the medium and A is the amplitude of the wave.

max energy density = $(1/2) * \rho * (0.2 \text{ meters})^2$

The density of air at standard temperature and pressure is approximately 1.2 kg/m^3.

max energy density = (1/2) * 1.2 kg/m^3 * (0.2 meters)^2 = 0.024 Joules/m^3.

Answer: the maximum energy density of the wave is 0.024 Joules/m^3.

e) A wave is traveling in a medium with a frequency of 300 Hz and a speed of 600 m/s. What is the wavelength of the wave?

$\lambda = v/f$, where λ is the wavelength of the wave, v is its speed, and f is its frequency.

λ = 600 m/s / 300 Hz = 2 meters.

Answer: the wavelength of the wave is 2 meters.

Exercise 3

Questions

a) A wave with a frequency of 500Hz and a wavelength of 0.2 meters is traveling in a medium with a speed of 100 m/s. What is the phase velocity of the wave?

b) A wave is traveling in a medium with a speed of 200 m/s and a frequency of 1000 Hz. What is the wavelength of the wave?

c) A wave is traveling in a medium with a wavelength of 0.5 meters and an amplitude of 0.3 meters. What is the maximum velocity of the particles in the medium?

d) A wave is traveling in a medium with a wavelength of 0.2 meters and a frequency of 1000 Hz. What is the period of the wave?

Answers

a) A wave with a frequency of 500Hz and a wavelength of 0.2 meters is traveling in a medium with a speed of 100 m/s. What is the phase velocity of the wave?
The phase velocity of a wave is the speed at which its phase is moving, and it is equal to the product of its frequency and wavelength.
The formula for the phase velocity of a wave is $v\_p = \lambda * f$, where v_p is the phase velocity, λ is the wavelength, and f is the frequency.
$v\_p = 0.2$ meters * 500 Hz = 100 m/s.
Answer: the phase velocity of the wave is 100 m/s.

b) A wave is traveling in a medium with a speed of 200 m/s and a frequency of 1000 Hz. What is the wavelength of the wave?
$\lambda = v/f$, where λ is the wavelength of the wave, v is its speed, and f is its frequency.
$\lambda = 200$ m/s / 1000 Hz = 0.2 meters.
Answer: the wavelength of the wave is 0.2 meters.

c) A wave is traveling in a medium with a wavelength of 0.5 meters and an amplitude of 0.3 meters. What is the maximum velocity of the particles in the medium?
The maximum velocity of the particles in a medium is equal to the product of the frequency and the amplitude of the wave.

The formula for the maximum velocity of the particles in a medium is:

vmax = 2πfA, where vmax is the maximum velocity of the particles, f is the frequency of the wave, and A is its amplitude.

vmax = 2π * 1/T * 0.3 meters, where T is the period of the wave.

The period of the wave can be calculated as T = λ/v, where λ is the wavelength of the wave and v is its speed.

T = 0.5 meters / 100 m/s = 0.005 seconds.

vmax = 2π * 1/0.005 seconds * 0.3 meters = 37.7 m/s.

Answer: the maximum velocity of the particles in the medium is 37.7 m/s.

d) A wave is traveling in a medium with a wavelength of 0.2 meters and a frequency of 1000 Hz. What is the period of the wave?

T = 1/f, where T is the period of the wave and f is its frequency.

T = 1/1000 Hz = 0.001 seconds.

Answer: the period of the wave is 0.001 seconds.

Exercise 4

Questions

a) A wave with a frequency of 100 Hz and an amplitude of 0.2 meters is traveling in a medium with a speed of 50 m/s. What is the maximum displacement of the particles in the medium?

b) Two waves with wavelengths of 0.5 meters and 0.4 meters, respectively, are traveling in the same medium. What is the beat frequency of the waves?

c) A wave is traveling in a medium with a frequency of 200 Hz and a wavelength of 0.1 meters. What is the period of the wave?

d) A wave is traveling in a medium with a speed of 300 m/s and a frequency of 1000 Hz. What is the wavelength of the wave?

e) A wave is traveling in a medium with a speed of 200 m/s and a wavelength of 0.4 meters. What is the frequency of the wave?

Answers

a) A wave with a frequency of 100 Hz and an amplitude of 0.2 meters is traveling in a medium with a speed of 50 m/s. What is the maximum displacement of the particles in the medium?

The maximum displacement of the particles in the medium can be found using the formula: maximum displacement = amplitude.

Answer: the maximum displacement of the particles in the medium is 0.2 meters.

b) Two waves with wavelengths of 0.5 meters and 0.4 meters, respectively, are traveling in the same medium. What is the beat frequency of the waves?

The beat frequency can be found using the formula: beat frequency = absolute value of (frequency of wave 1 - frequency of wave 2).

Answer: the beat frequency is |0.5 - 0.4| = 0.1 Hz.

c) A wave is traveling in a medium with a frequency of 200 Hz and a wavelength of 0.1 meters. What is the period of the wave?

The period of the wave can be found using the formula: period = 1 / frequency.

Answer: the period of the wave is 1 / 200 = 0.005 seconds.

d) A wave is traveling in a medium with a speed of 300 m/s and a frequency of 1000 Hz. What is the wavelength of the wave?

wavelength = speed / frequency.

Answer: the wavelength of the wave is 300 / 1000 = 0.3 meters.

e) A wave is traveling in a medium with a speed of 200 m/s and a wavelength of 0.4 meters. What is the frequency of the wave?

frequency = speed / wavelength.

Answer: the frequency of the wave is 200 / 0.4 = 500 Hz.

Exercise 5

Questions

a) A wave is traveling in a medium with a wavelength of 0.1 meters and a period of 0.01 seconds. What is the speed of the wave?

b) A wave is traveling in a medium with a speed of 100 m/s and a wavelength of 0.5 meters. What is the frequency of the wave?

c) A wave is traveling in a medium with a wavelength of 0.2 meters and a frequency of 500 Hz. What is the period of the wave?

d) A wave is traveling in a medium with an amplitude of 0.3 meters and a maximum velocity of 50 m/s. What is the frequency of the wave?

e) A wave is traveling in a medium with a speed of 200 m/s and a wavelength of 0.5 meters. What is the period of the wave?

Answers

a) A wave is traveling in a medium with a wavelength of 0.1 meters and a period of 0.01 seconds. What is the speed of the wave?

speed = wavelength / period.

Answer: the speed of the wave is 0.1 / 0.01 = 10 m/s.

b) A wave is traveling in a medium with a speed of 100 m/s and a wavelength of 0.5 meters. What is the frequency of the wave?

frequency = speed / wavelength.

Answer: the frequency of the wave is 100 / 0.5 = 200 Hz.

c) A wave is traveling in a medium with a wavelength of 0.2 meters and a frequency of 500 Hz. What is the period of the wave?

period = 1 / frequency.

Answer: the period of the wave is 1 / 500 = 0.002 seconds.

d) A wave is traveling in a medium with an amplitude of 0.3 meters and a maximum velocity of 50 m/s. What is the frequency of the wave?

frequency = maximum velocity / wavelength.

Answer: the frequency of the wave is 50 / 0.3 = 166.7 Hz.

e) A wave is traveling in a medium with a speed of 200 m/s and a wavelength of 0.5 meters. What is the period of the wave?

period = wavelength / speed.

Answer: the period of the wave is 0.5 / 200 = 0.0025 seconds.

Part 5: Waves

Exercise 1

Questions

a) A wave with a frequency of 500 Hz and a wavelength of 0.2 meters is traveling in a medium. What is the wave speed?

wave speed = frequency x wavelength.

Answer: the wave speed is 500 x 0.2 = 100 m/s.

b) A wave is traveling in a medium with a wavelength of 0.5 meters and a frequency of 200 Hz. What is the wave speed?

wave speed = frequency x wavelength.

Answer: the wave speed is 200 x 0.5 = 100 m/s.

c) A wave with an amplitude of 0.3 meters and a frequency of 100 Hz is traveling in a medium with a speed of 50 m/s. What is the maximum velocity of the particles in the medium?

maximum velocity = amplitude x frequency x 2π.

Answer: the maximum velocity of the particles in the medium is 0.3 x 100 x 2π = 18.85 m/s.

d) Two waves with amplitudes of 0.2 meters and 0.3 meters, respectively, are traveling in the same medium. What is the amplitude of the resulting wave?

amplitude = √(amplitude of wave 1)^2 + (amplitude of wave 2)^2.

Answer: the amplitude of the resulting wave is √(0.2^2 + 0.3^2) = 0.36 meters.

e) A wave is traveling in a medium with a wavelength of 0.1 meters and a speed of 200 m/s. What is the frequency of the wave?

frequency = speed / wavelength.

Answer: the frequency of the wave is 200 / 0.1 = 2000 Hz.

Answers

a) A wave with a frequency of 500 Hz and a wavelength of 0.2 meters is traveling in a medium. What is the wave speed?

wave speed = frequency x wavelength.

Answer: the wave speed is 500 x 0.2 = 100 m/s.

b) A wave is traveling in a medium with a wavelength of 0.5 meters and a frequency of 200 Hz. What is the wave speed?

wave speed = frequency x wavelength.

Answer: the wave speed is 200 x 0.5 = 100 m/s.

c) A wave with an amplitude of 0.3 meters and a frequency of 100 Hz is traveling in a medium with a speed of 50 m/s. What is the maximum velocity of the particles in the medium?

maximum velocity = amplitude x frequency x 2π.

Answer: the maximum velocity of the particles in the medium is 0.3 x 100 x 2π = 18.85 m/s.

d) Two waves with amplitudes of 0.2 meters and 0.3 meters, respectively, are traveling in the same medium. What is the amplitude of the resulting wave?

amplitude = √(amplitude of wave 1)^2 + (amplitude of wave 2)^2.

Answer: the amplitude of the resulting wave is √(0.2^2 + 0.3^2) = 0.36 meters.

e) A wave is traveling in a medium with a wavelength of 0.1 meters and a speed of 200 m/s. What is the frequency of the wave?

frequency = speed / wavelength.

Answer: the frequency of the wave is 200 / 0.1 = 2000 Hz.

Exercise 2

Questions

a) A wave is traveling in a medium with a speed of 300 m/s and a wavelength of 0.3 meters. What is the period of the wave?

b) A wave is traveling in a medium with a speed of 100 m/s and a frequency of 50 Hz. What is the wavelength of the wave?

c) A wave is traveling in a medium with a wavelength of 0.02 meters and a speed of 500 m/s. What is the frequency of the wave?

d) A wave with an amplitude of 0.4 meters and a wavelength of 0.5 meters is traveling in a medium. What is the maximum displacement of the particles in the medium?

e) A wave is traveling in a medium with a frequency of 200 Hz and a period of 0.005 seconds. What is the wavelength of the wave?

Answers

a) A wave is traveling in a medium with a speed of 300 m/s and a wavelength of 0.3 meters. What is the period of the wave?

period = wavelength / speed.

Answer: the period of the wave is 0.3 / 300 = 0.001 seconds.

b) A wave is traveling in a medium with a speed of 100 m/s and a frequency of 50 Hz. What is the wavelength of the wave?

wavelength = speed / frequency.

Answer: the wavelength of the wave is 100 / 50 = 2 meters.

c) A wave is traveling in a medium with a wavelength of 0.02 meters and a speed of 500 m/s. What is the frequency of the wave?

frequency = speed / wavelength.

Answer: the frequency of the wave is 500 / 0.02 = 25,000 Hz.

d) A wave with an amplitude of 0.4 meters and a wavelength of 0.5 meters is traveling in a medium. What is the maximum displacement of the particles in the medium?

maximum displacement = amplitude.

Answer: the maximum displacement of the particles in the medium is 0.4 meters.

e) A wave is traveling in a medium with a frequency of 200 Hz and a period of 0.005 seconds. What is the wavelength of the wave?

wavelength = speed x period.

Answer: the wavelength of the wave is 1000 x 0.005 = 5 meters.

Exercise 3

Questions

a) A wave is traveling in a medium with a speed of 100 m/s and a wavelength of 0.5 meters. What is the frequency of the wave?

b) A wave is traveling in a medium with a frequency of 100 Hz and a period of 0.01 seconds. What is the wavelength of the wave?

c) A wave is traveling in a medium with an amplitude of 0.5 meters and a maximum displacement of 0.7 meters. What is the frequency of the wave?

d) A wave with a frequency of 300 Hz and a wavelength of 0.2 meters is traveling in a medium. What is the wave speed?

e) A wave is traveling in a medium with a speed of 150 m/s and a wavelength of 0.4 meters. What is the frequency of the wave?

Answers

a) A wave is traveling in a medium with a speed of 100 m/s and a wavelength of 0.5 meters. What is the frequency of the wave?

frequency = speed / wavelength.

Answer: the frequency of the wave is 100 / 0.5 = 200 Hz.

b) A wave is traveling in a medium with a frequency of 100 Hz and a period of 0.01 seconds. What is the wavelength of the wave?

wavelength = speed x period

Answer: the wavelength of the wave is 1 meter.

c) A wave is traveling in a medium with an amplitude of 0.5 meters and a maximum displacement of 0.7 meters. What is the frequency of the wave?

frequency = maximum displacement / amplitude.

Answer: the frequency of the wave is 0.7 / 0.5 = 1.4 Hz.

d) A wave with a frequency of 300 Hz and a wavelength of 0.2 meters is traveling in a medium. What is the wave speed?

wave speed = frequency x wavelength.

Answer: the wave speed is 300 x 0.2 = 60 m/s.

e) A wave is traveling in a medium with a speed of 150 m/s and a wavelength of 0.4 meters. What is the frequency of the wave?

frequency = speed / wavelength

Answer: the frequency of the wave is 150 / 0.4 = 375 Hz.

Exercise 4

Questions

a) A wave is traveling in a medium with a wavelength of 0.1 meters and a frequency of 50 Hz. What is the period of the wave?

b) A wave with a frequency of 500 Hz and a wavelength of 0.01 meters is traveling in a medium. What is the wave speed?

c) A wave is traveling in a medium with a wavelength of 0.3 meters and a frequency of 100 Hz. What is the wave speed?

d) A wave with a wavelength of 0.05 meters is traveling in a medium with a frequency of 400 Hz. What is the wave speed?

e) A wave is traveling in a medium with a speed of 200 m/s and a frequency of 50 Hz. What is the wavelength of the wave?

Answers

a) A wave is traveling in a medium with a wavelength of 0.1 meters and a frequency of 50 Hz. What is the period of the wave?
period = 1 / frequency
Answer: the period of the wave is 1 / 50 = 0.02 seconds.

b) A wave with a frequency of 500 Hz and a wavelength of 0.01 meters is traveling in a medium. What is the wave speed?
wave speed = frequency x wavelength
Answer: the wave speed is 500 x 0.01 = 5 m/s.

c) A wave is traveling in a medium with a wavelength of 0.3 meters and a frequency of 100 Hz. What is the wave speed?
wave speed = frequency x wavelength
Answer: the wave speed is 30 m/s.

d) A wave with a wavelength of 0.05 meters is traveling in a medium with a frequency of 400 Hz. What is the wave speed?
wave speed = frequency x wavelength
Answer: the wave speed is 20 m/s.

e) A wave is traveling in a medium with a speed of 200 m/s and a frequency of 50 Hz. What is the wavelength of the wave?
wavelength = speed / frequency

Answer: the wavelength of the wave is 200 / 50 = 4 meters.

Exercise 1

Questions

a) A wave with a wavelength of 0.02 meters and a frequency of 1000 Hz is traveling in a medium. What is the wave speed?

b) A wave is traveling in a medium with a frequency of 200 Hz and a wavelength of 0.4 meters. What is the period of the wave?

c) A wave with a wavelength of 0.1 meters and a period of 0.02 seconds is traveling in a medium. What is the wave speed?

d) A wave is traveling in a medium with a frequency of 1000 Hz and a wavelength of 0.001 meters. What is the wave speed?

e) A wave is traveling in a medium with an amplitude of 0.2 meters and a maximum velocity of 5 m/s. What is the wavelength of the wave?

Answers

a) A wave with a wavelength of 0.02 meters and a frequency of 1000 Hz is traveling in a medium. What is the wave speed?

wave speed = frequency x wavelength.

Answer: the wave speed is 20 m/s.

b) A wave is traveling in a medium with a frequency of 200 Hz and a wavelength of 0.4 meters. What is the period of the wave?

period = 1 / frequency

Answer: the period of the wave is 1 / 200 = 0.005 seconds.

c) A wave with a wavelength of 0.1 meters and a period of 0.02 seconds is traveling in a medium. What is the wave speed?

wave speed = wavelength / period.

Answer: the wave speed is 5 m/s.

d) A wave is traveling in a medium with a frequency of 1000 Hz and a wavelength of 0.001 meters. What is the wave speed?

wave speed = frequency x wavelength

Answer: the wave speed is 1 m/s.

e) A wave is traveling in a medium with an amplitude of 0.2 meters and a maximum velocity of 5 m/s. What is the wavelength of the wave?

wavelength = maximum velocity / (amplitude x 2π).

Answer: the wavelength of the wave is 5 / (0.2 x 2π) = 3.98 meters

Exercise 2

Questions

a) A wave with a frequency of 300 Hz and a period of 0.005 seconds is traveling in a medium. What is the wavelength of the wave?

b) A wave with a wavelength of 0.04 meters and a wave speed of 40 m/s is traveling in a medium. What is the frequency of the wave?

c) A wave is traveling in a medium with a wavelength of 0.2 meters and a wave speed of 400 m/s. What is the frequency of the wave?

d) A wave with a frequency of 100 Hz and a wave speed of 10 m/s is traveling in a medium. What is the wavelength of the wave?

e) A wave is traveling in a medium with a wavelength of 0.1 meters and a frequency of 200 Hz. What is the wave speed?

Answers

a) A wave with a frequency of 300 Hz and a period of 0.005 seconds is traveling in a medium. What is the wavelength of the wave?

wavelength = wave speed / frequency

Answer: the wavelength of the wave is 50 / 300 = 0.167 meters.

b) A wave with a wavelength of 0.04 meters and a wave speed of 40 m/s is traveling in a medium. What is the frequency of the wave?

frequency = wave speed / wavelength

Answer: the frequency of the wave is 40 / 0.04 = 1000 Hz.

c) A wave is traveling in a medium with a wavelength of 0.2 meters and a wave speed of 400 m/s. What is the frequency of the wave?

frequency = wave speed / wavelength

Answer: the frequency of the wave is 400 / 0.2 = 2000 Hz.

d) A wave with a frequency of 100 Hz and a wave speed of 10 m/s is traveling in a medium. What is the wavelength of the wave?

wavelength = wave speed / frequency

Answer: the wavelength of the wave is 10 / 100 = 0.1 meters.

e) A wave is traveling in a medium with a wavelength of 0.1 meters and a frequency of 200 Hz. What is the wave speed?

wave speed = frequency x wavelength

Answer: the wave speed is 20 m/s.

Exercise 3

Questions

a) A wave with a wavelength of 0.01 meters and a wave speed of 300 m/s is traveling in a medium. What is the frequency of the wave?

b) A wave is traveling in a medium with a frequency of 400 Hz and a wavelength of 0.1 meters. What is the wave speed?

c) A wave with a wavelength of 0.02 meters is traveling in a medium with a frequency of 500 Hz. What is the wave speed?

d) A wave is traveling in a medium with a wavelength of 0.5 meters and a wave speed of 50 m/s. What is the frequency of the wave?

e) A wave with a frequency of 1000 Hz and a wavelength of 0.002 meters is traveling in a medium. What is the wave speed?

Answers

a) A wave with a wavelength of 0.01 meters and a wave speed of 300 m/s is traveling in a medium. What is the frequency of the wave?

frequency = wave speed / wavelength

Answer: the frequency of the wave is 300 / 0.01 = 30,000 Hz.

b) A wave is traveling in a medium with a frequency of 400 Hz and a wavelength of 0.1 meters. What is the wave speed?

wave speed = frequency x wavelength

Answer: the wave speed is 40 m/s.

c) A wave with a wavelength of 0.02 meters is traveling in a medium with a frequency of 500 Hz. What is the wave speed?

wave speed = frequency x wavelength

Answer: the wave speed is 10 m/s.

d) A wave is traveling in a medium with a wavelength of 0.5 meters and a wave speed of 50 m/s. What is the frequency of the wave?

frequency = wave speed / wavelength

Answer: the frequency of the wave is 50 / 0.5 = 100 Hz.

e) A wave with a frequency of 1000 Hz and a wavelength of 0.002 meters is traveling in a medium. What is the wave speed?

wave speed = frequency x wavelength

Answer: the wave speed is 2 m/s

Exercise 4

Questions

a) A wave is traveling in a medium with an amplitude of 0.1 meters and a wavelength of 0.02 meters. What is the maximum displacement of the wave?

b) A wave is traveling in a medium with a frequency of 1000 Hz and an amplitude of 0.2 meters. What is the maximum speed of the particles in the medium?
140.

c) A wave with a wavelength of 0.05 meters is traveling in a medium with a frequency of 100 Hz. What is the phase difference between two particles that are 0.02 meters apart?

d) A wave is traveling in a medium with a wavelength of 0.1 meters and a frequency of 200 Hz. What is the phase difference between two particles that are 0.03 meters apart?

e) A wave with a frequency of 500 Hz and a wavelength of 0.02 meters is traveling in a medium. What is the period of the wave?

Answers

a) A wave is traveling in a medium with an amplitude of 0.1 meters and a wavelength of 0.02 meters. What is the maximum displacement of the wave?

maximum displacement = amplitude x 2

Answer: the maximum displacement of the wave is 0.2 meters.

b) A wave is traveling in a medium with a frequency of 1000 Hz and an amplitude of 0.2 meters. What is the maximum speed of the particles in the medium?

140.

maximum speed = 2π x frequency x amplitude

Answer: the maximum speed of the particles is 2π x 1000 x 0.2 = 1256.64 m/s.

c) A wave with a wavelength of 0.05 meters is traveling in a medium with a frequency of 100 Hz. What is the phase difference between two particles that are 0.02 meters apart?

phase difference = (2π / wavelength) x distance

Answer: the phase difference between the two particles is (2π / 0.05) x 0.02 = 2.513 radians.

d) A wave is traveling in a medium with a wavelength of 0.1 meters and a frequency of 200 Hz. What is the phase difference between two particles that are 0.03 meters apart?

phase difference = (2π / wavelength) x distance

Answer: the phase difference between the two particles is (2π / 0.1) x 0.03 = 1.884 radians.

e) A wave with a frequency of 500 Hz and a wavelength of 0.02 meters is traveling in a medium. What is the period of the wave?

period = 1 / frequency

Answer: the period of the wave is 1 / 500 = 0.002 seconds.

Exercise 5

Questions

a) A wave is traveling in a medium with a period of 0.01 seconds and an amplitude of 0.1 meters. What is the maximum velocity of the particles in the medium?

b) A wave with a frequency of 100 Hz and an amplitude of 0.2 meters is traveling in a medium. What is the energy density of the wave?

c) A wave is traveling in a medium with an energy density of 0.2 J/m^3 and a wavelength of 0.1 meters. What is the amplitude of the wave?

d) A wave with a frequency of 200 Hz and a wavelength of 0.02 meters is traveling in a medium. What is the phase velocity of the wave?

e) A wave is traveling in a medium with a wavelength of 0.05 meters and a period of 0.01 seconds. What is the wave speed?

Answers

a) A wave is traveling in a medium with a period of 0.01 seconds and an amplitude of 0.1 meters. What is the maximum velocity of the particles in the medium?

maximum velocity = 2π x amplitude / period

Answer: the maximum velocity of the particles is 2π x 0.1 / 0.01 = 6.28 m/s.

b) A wave with a frequency of 100 Hz and an amplitude of 0.2 meters is traveling in a medium. What is the energy density of the wave?

energy density = 0.5 x density x wave speed^2 x amplitude^2

Answer: the energy density of the wave is 0.5 x 1 x (2π x 100 x 0.2)^2 = 125.66 J/m^3.

c) A wave is traveling in a medium with an energy density of 0.2 J/m^3 and a wavelength of 0.1 meters. What is the amplitude of the wave?

amplitude = √(2 x energy density / density x wave speed^2) = √(2 x 0.2 / 1 x (2π x 200))^2 = 0.002 m.

d) A wave with a frequency of 200 Hz and a wavelength of 0.02 meters is traveling in a medium. What is the phase velocity of the wave?

phase velocity = frequency x wavelength

Answer: the phase velocity of the wave is 200 x 0.02 = 4 m/s.

e) A wave is traveling in a medium with a wavelength of 0.05 meters and a period of 0.01 seconds. What is the wave speed?

wave speed = wavelength / period

Answer: the wave speed is 0.05 / 0.01 = 5 m/s.

Exercise 1

Questions

a) A wave with a frequency of 1000 Hz and a wavelength of 0.01 meters is traveling in a medium. What is the phase difference between two particles that are 0.05 meters apart?

b) An electromagnetic wave has a frequency of 5.0×10^{14} Hz and a wavelength of 600 nm. What is the speed of the wave in a vacuum?

c) A wave has a period of 0.02 seconds and a wavelength of 2 meters. What is the speed of the wave?

d) A wave travels through a medium with a speed of 50 m/s and a frequency of 100 Hz. What is the wavelength of the wave?

e) A wave with a frequency of 50 Hz and a wavelength of 10 meters travels through a medium with a speed of 500 m/s. What is the period of the wave?

Answers

a) A wave with a frequency of 1000 Hz and a wavelength of 0.01 meters is traveling in a medium. What is the phase difference between two particles that are 0.05 meters apart?

The phase difference = (2π / wavelength) x distance

Answer: the phase difference between the two particles is (2π / 0.01) x 0.05 = 3.14 radians.

b) An electromagnetic wave has a frequency of 5.0×10^{14} Hz and a wavelength of 600 nm. What is the speed of the wave in a vacuum?

speed = frequency x wavelength.

speed = (5.0×10^{14} Hz) x (600×10^{-9} m) = 3.0×10^{8} m/s.

Answer: the speed of the wave in a vacuum is 3.0×10^{8} m/s.

c) A wave has a period of 0.02 seconds and a wavelength of 2 meters. What is the speed of the wave?

speed = wavelength / period

speed = 2 m / 0.02 s = 100 m/s

Answer: the speed of the wave is 100 m/s.

d) A wave travels through a medium with a speed of 50 m/s and a frequency of 100 Hz. What is the wavelength of the wave?

wavelength = speed / frequency

wavelength = 50 m/s / 100 Hz = 0.5 m

Answer: the wavelength of the wave is 0.5 m.

e) A wave with a frequency of 50 Hz and a wavelength of 10 meters travels through a medium with a speed of 500 m/s. What is the period of the wave?

speed = wavelength x frequency

wavelength = speed / frequency

wavelength = 500 m/s / 50 Hz = 10 m

Answer: the wavelength of the wave is 10 m

Exercise 2

Questions

a) A wave has a frequency of 200 Hz and a velocity of 400 m/s. What is the wavelength of the wave?

b) A string has a mass of 0.05 kg and is stretched between two fixed points that are 1 meter apart. If a wave with a wavelength of 2 meters travels on the string, what is the speed of the wave?

c) A wave with a wavelength of 0.5 meters travels through a medium with a speed of 150 m/s. What is the frequency of the wave?

d) A sound wave with a frequency of 500 Hz travels through a medium with a velocity of 340 m/s. What is the wavelength of the wave?

e) A wave with a wavelength of 4 meters and a frequency of 50 Hz travels through a medium with a speed of 200 m/s. What is the period of the wave?

Answers

a) A wave has a frequency of 200 Hz and a velocity of 400 m/s. What is the wavelength of the wave?

velocity = frequency x wavelength.

wavelength = velocity / frequency.

wavelength = 400 m/s / 200 Hz = 2 m.

Answer: the wavelength of the wave is 2 m.

b) A string has a mass of 0.05 kg and is stretched between two fixed points that are 1 meter apart. If a wave with a wavelength of 2 meters travels on the string, what is the speed of the wave?

speed = square root(tension / linear density).

speed = square root(1.0 N / 0.05 kg) = 4.47 m/s.

Answer: the speed of the wave is 4.47 m/s.

c) A wave with a wavelength of 0.5 meters travels through a medium with a speed of 150 m/s. What is the frequency of the wave?

frequency = speed / wavelength.

wavelength = speed / frequency

wavelength = 150 m/s / 0.5 m = 300 Hz.

Answer: the frequency of the wave is 300 Hz.

d) A sound wave with a frequency of 500 Hz travels through a medium with a velocity of 340 m/s. What is the wavelength of the wave?

wavelength = velocity / frequency.

wavelength = 340 m/s / 500 Hz = 0.68 m.

Answer: the wavelength of the wave is 0.68 m.

e) A wave with a wavelength of 4 meters and a frequency of 50 Hz travels through a medium with a speed of 200 m/s. What is the period of the wave?

period = 1 / frequency

period = 1 / 50 Hz = 0.02 s

Answer: the period of the wave is 0.02 s.

Exercise 3

Questions

a) A wave with a frequency of 100 Hz has a velocity of 200 m/s. What is the wavelength of the wave?

b) A transverse wave is traveling on a rope with a frequency of 10 Hz and a wavelength of 2 meters. What is the speed of the wave?

c) A sound wave has a wavelength of 0.5 meters and a frequency of 440 Hz. What is the speed of the wave?

d) A wave is traveling on a string with a frequency of 5 Hz and an amplitude of 0.1 meters. What is the maximum displacement of the particles in the string?

e) A longitudinal wave is traveling through a material with a speed of 1500 m/s and a frequency of 100 Hz. What is the wavelength of the wave?

Answers

a) A wave with a frequency of 100 Hz has a velocity of 200 m/s. What is the wavelength of the wave?

wavelength = speed / frequency.

wavelength = 200 m/s / 100 Hz = 2 m.

Answer: the wavelength of the wave is 2 m.

b) A transverse wave is traveling on a rope with a frequency of 10 Hz and a wavelength of 2 meters. What is the speed of the wave?

$v = f\lambda$ to solve this problem, where v is the speed of the wave, f is the frequency, and λ is the wavelength.

$v = (10 \text{ Hz})(2 \text{ m}) = 20 \text{ m/s}$.

Answer: the speed of the wave is 20 m/s.

c) A sound wave has a wavelength of 0.5 meters and a frequency of 440 Hz. What is the speed of the wave?

$v = f\lambda$, but this time we need to use the speed of sound, which is approximately 343 m/s at room temperature.

$v = (440 \text{ Hz})(0.5 \text{ m}) = 220 \text{ m/s}$.

Answer: the speed of the wave is 220 m/s.

d) A wave is traveling on a string with a frequency of 5 Hz and an amplitude of 0.1 meters. What is the maximum displacement of the particles in the string?

The maximum displacement of the particles in a wave is equal to the amplitude of the wave.

Answer: the maximum displacement in this case is 0.1 meters.

e) A longitudinal wave is traveling through a material with a speed of 1500 m/s and a frequency of 100 Hz. What is the wavelength of the wave?

$v = f\lambda$

$\lambda = v/f$

$\lambda = 1500$ m/s / 100 Hz = 15 meters.

Answer: the wavelength of the wave is 15 meters.

Exercise 4

Questions

a) A wave is traveling on a string with a frequency of 8 Hz and a wavelength of 0.5 meters. What is the period of the wave?

b) Two waves are traveling on a string in opposite directions. One wave has a frequency of 10 Hz and an amplitude of 0.2 meters, while the other wave has a frequency of 15 Hz and an amplitude of 0.1 meters. What is the amplitude of the resulting wave?

c) A guitar string has a length of 60 cm and a mass of 3 g. If the string is under a tension of 500 N, what is the speed of a wave on the string?

d) A sound wave of frequency 440 Hz is traveling through air with a speed of 340 m/s. What is the wavelength of the sound wave?

e) A water wave has a wavelength of 3 m and a frequency of 2 Hz. What is the speed of the wave?

Answers

a) A wave is traveling on a string with a frequency of 8 Hz and a wavelength of 0.5 meters. What is the period of the wave?

$T = 1/f$ to solve for the period, where T is the period and f is the frequency.

$T = 1/8$ Hz = 0.125 seconds.

Answer: the period of the wave is 0.125 seconds.

b) Two waves are traveling on a string in opposite directions. One wave has a frequency of 10 Hz and an amplitude of 0.2 meters, while the other wave has a frequency of 15 Hz and an amplitude of 0.1 meters. What is the amplitude of the resulting wave?

When two waves interfere with each other, the displacement of the resulting wave at any given point is the sum of the displacements of the individual waves at that point.

The amplitude of the resulting wave is equal to the sum of the amplitudes of the individual waves.

In this case, the amplitude of the resulting wave is 0.2 meters + 0.1 meters = 0.3 meters.

Answer: the amplitude of the resulting wave is 0.3 meters.

c) A guitar string has a length of 60 cm and a mass of 3 g. If the string is under a tension of 500 N, what is the speed of a wave on the string?

$v = sqrt(T/u)$, where T is the tension in the string and u is the linear mass density of the string.

$u = m/L$, where m is the mass of the string and L is the length of the string.

$u = 3$ g / 60 cm = 0.05 g/cm = 0.0005 kg/m.

Answer: $v = sqrt(500$ N / 0.0005 kg/m$) = 100$ m/s.

d) A sound wave of frequency 440 Hz is traveling through air with a speed of 340 m/s. What is the wavelength of the sound wave?

$v = f\lambda$, where f is the frequency of the sound wave and λ is the wavelength.

Answer: $\lambda = v/f = 340$ m/s / 440 Hz = 0.773 m.

e) A water wave has a wavelength of 3 m and a frequency of 2 Hz. What is the speed of the wave?

$v = f\lambda$, where f is the frequency of the wave and λ is the wavelength.

Answer: $v = 2$ Hz x 3 m = 6 m/s.

Exercise 5

Questions

a) A guitar string is plucked, producing a wave with a wavelength of 40 cm and a frequency of 500 Hz. What is the speed of the wave?

b) A wave on a string has an amplitude of 0.5 cm and a wavelength of 20 cm. What is the maximum displacement of a particle on the string?

Answers

a) A guitar string is plucked, producing a wave with a wavelength of 40 cm and a frequency of 500 Hz. What is the speed of the wave?

$v = f\lambda$, where f is the frequency of the wave and λ is the wavelength.

Answer: v = 500 Hz x 0.4 m = 200 m/s.

b) A wave on a string has an amplitude of 0.5 cm and a wavelength of 20 cm. What is the maximum displacement of a particle on the string?

Answer: the maximum displacement of a particle on a string is equal to the amplitude of the wave = 0.5 cm.

Conclusion

Thank you once again for purchasing this book. I hope it has helped you in your journey to understand the basics of waves.

Please, if you learnt something from this book, I would like you to leave a review. It'd be appreciated.

Thank you.

www.ingramcontent.com/pod-product-compliance
Lightning Source LLC
Chambersburg PA
CBHW080907220526

45466CB00011BA/3499